A Complete

Guide of Palmistry

Written and Edited by

Ravi R Naik

About Author

Koto the pen name of **Ravi R Naik** is a suspense novel writer, film writer part time blogger as well as astrologer born on 28 January 1995 in south India. He started his writings in his mother tongue Telugu at the age of five. The NEW D-street is the first novel translated by him in English. He written more than 25 novels in every category. He is a full time writer as well as movie story writer. He spend most of his time writing. He is a student studying under graduate. This book is dedicated to all his friends.

Meet him online

www.ask-astrologer.blogspot.com

Follow him on twitter @ravinaikwriter

Foreword

All the information given in this Palmistry guide is for reference purpose only and not to hurt anyone. All the articles and palmistry examples are given after the keen experience and not copied from any other sources. Hope you understand. All the article part is for interesting purpose only. All the information is very easy to understand. This book will help basic astrologers who are trying to learn palmistry. All articles are copyrighted to our site and book. We are not responsible for any content in this. All the information is given out of reviews and experience.

-Author

COMPLETE GUIDE OF PALMISTRY

After my seven years experience I have found many mysterious information about Palmistry. Before I explain the complete study of palmistry you must believe in it. After my research on hundreds of people I came to know many things about palmistry. People actually believe that palmistry is blind belief. But I will say that palmistry is a science explaining the perfect guidance of our future. Most scientists now believe that palmistry works. Actually there is a lot of difference between palmistry and astrology. In order to explain each and every line of palmistry you must first know how lines are formed. The major lines head line, heart line and life line are formed at the birth, this is of reason that the child will hold his fist close together which makes the responsible for the marks. That marks are not changes in his whole life. The three marks define the person's character, health, intelligent and life span.

According to me the main reason for forming of lines other than the three major lines are the blood circulating in our body will sometimes slows at particular position in palm which makes the line appears. On the other hand, the lines are formed because of our nerves interactions.

Now let me directly enter the study of palmistry. In these chapters am going to explain each and every line in our palm. Before you start examine the lines you need to know few things. Which I will explain as question and answers.

1. Is palmistry true?

   A : Yes. Because I have examined hundreds of palms of poor, rich, business person, artist, beggar and millionaire.
   You can believe in palmistry. We already knew Cheiro's palmistry, but my palmistry found many changes than Cheiro's. with the help of my palmistry you can find the every persons behavior, character, intelligent, luck, love, happiness, earnings and even crime.

Not only the palmistry I will explain some mysterious articles of our hand.
All the articles in this book I have explained in a easy way so that one can understand easily.

2. What about Planets and Sun signs?
A: Some people believe in Planets rather than Palmistry. Most Scientists and Astrologer found that the Planets have no influence in our life. But I will give a small information which makes you believe in Planets influence.
Sun, Moon, Mercury, Venus, Mars, Jupiter, Saturn, Neptune, Pluto are the 9 Planets that shows influence in our life.

Let us take an example of Sun. We already know without sun there is nothing on Earth. For example let us take two persons, one from India and other from America. We can easily identify the person just seeing at them, because of their color, speech and body language. Let us take a person is having too much sun bath. We can say that the person will likely to affect from some diseases. According to science taking excess of sun bath daily will lead to dangerous diseases like Cancer. This is a direct research, but the indirect thing here is the person affects from Cancer only if he is likely to have too much sun bath. This will clearly show that there is influence of sun in our life.

For example if you see the Moon you feel like relaxing, stress less and peace. Here moon has some influence on us which makes us relax or other..... This also proves that Moon has some influence on us. In the same way the planets shows their effect on human beings which are responsible for rich, poor, good, bad, happy, unhappy, luck, unlucky and so on...

3. What about Horoscopes?

A: Most of my friends, Palmistry seekers will ask me whether the horoscopes will have influence on our future.

The answer is No. Most of astrologer explains daily horoscopes saying this year is good, this year is bad and blablabla.... Actually no horoscope is true. They do not show effect on our future. They are for nothing. My kind information is please do not believe in horoscopes. The sun signs are just an identification marks which helps one person to know which planet is influencing him.

One more important information is that the planets will not show their effects unnecessarily, for example a person is born under the influence of Venus, this person will be happy, rich, prosperous. All his planets influences are defined first itself, they again do not change.

4. What about moles will they have influence in our future?

A: Yes. They have, let us take an example..
A person is having a mole on his knees, it
explains that the person will be rich
because of his wife's influence. You can
take the moles as planets that influence at
that particular position. But this is not the
perfect time to explain about this.

5. How can we find a particular planet is
influencing?
A: Most of us do not know which Planets
is influencing. The answer is simple; I will
explain how to identify the planets
influences in our life. In palmistry the
Planets are called Mounts. I will explain
what the particular mount signifies.

MOUNTS AND THEIR EFFECTS:

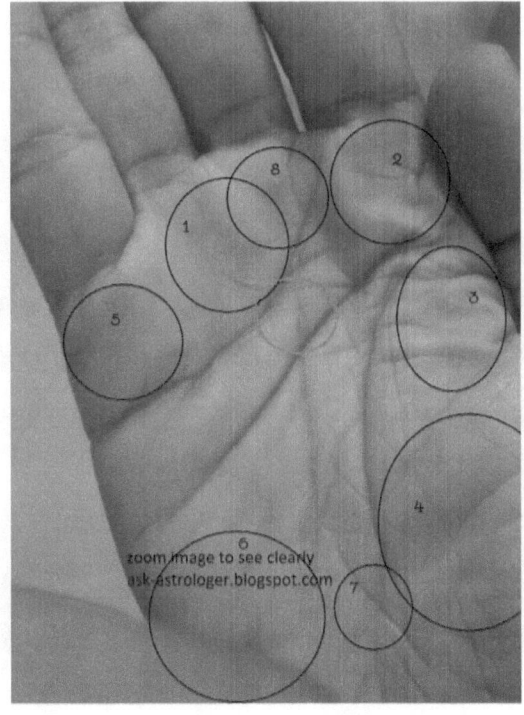

1. Sun Mount
2. Jupiter Mount
3. Mars Mount
4. Venus Mount
5. Mercury Mount
6. Moon Mount
7. Neptune Mount
8. Saturn Mount

Sun Mount: This one of the most important mount to have. If no sun mount no sun light to our future. The persons born under this mount will become familiar to society very quickly. Politicians, Leaders will come under this category.

Moon Mount : this mount will have a good influence in one's life. A person who is influenced by this planets can have his mount elevated. The persons having this mount will be rich, peace loving, travelers, lovers and artists.

Venus Mount : this mount will have a good influence in one's life. A person who is influence by this planet can have his mount elevated. The persons having this mount will be Artists; Will have prosperous, romantic, kind hearted as well as philosophic character. Most of Film Artists, Big Business persons, Teachers comes under this category.

Mars Mount : This mount will have some good as well as some negative influence. The person having mars planet influencing will be Anger, Short temper and Hard Workers. Software engineers, Private workers, team leaders will come under this category.

Jupiter Mount : One of the most powerful Mount to have. The people of this mount will be straight forward, Confident, intelligent, caretakers and dignified. All major big categories like a CEO of a Company, Administrator, IAS, IPS or a person of high position will have this mount. These peoples look so simple, Naughty, Childlike but are really great intelligent.

Saturn Mount : This is also one of the most powerful mount to have. The people of this mount will be hardworking, ignorance, philosophers, and limitation maintainers. A CEO of multiple brands, Brand Ambassadors, Scientists, and Millionaires will come under this category.

Mercury Mount : This is the smallest planet which shows their influence in one's life. The people of this mount will be timid, Hard workers, Helpers and good listeners. They may neither rich nor poor. They are simple peoples. Mercury influence people are the one who can save money for

further purpose. Govt employees, Super
market workers, computer operators,
factory workers will come under this
category.

Neptune Mount : This is the mount one
should not have. Neptune is so far from
earth. it may not show its influence on
people who born under other planets. If a
person is born under this planet's
influence, the person will be jobless,
criminals, terrorists or any bad posts.

Pluto Mount: The planet will not even have
their influence in one's life. It is too far to
look after their influencing one. No person
will be influenced by this. The children
who born under this planet's influence
will die before at the age of 5 because the
planet does not show its effect. So the
children who die under age of 5 or who
die at birth mostly likely because of this
planet.

I have not explained everything in detail. I just gave a simple reference to understand. This is not the time to explain complete mounts.

In this Article Am going to explain every mark on our palm and what it signifies.

HEART LINE, HEAD LINE AND LIFE LINE

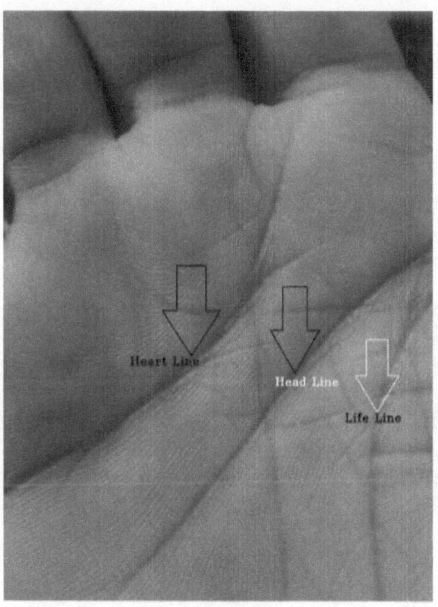

Fig : 1.1

HEART LINE

Heart Line : One of the most important line in Palm. A heart line signifies ones character and health problems.

We can easily identify one's character seeing at heart line.

A heart line with twists and curves signifies that the person is innocent and timid.

A heart line starting from Jupiter Mount signifies that the person is kind, good hearted, naughty and intelligent. He/she will have friendly nature.

A heart line starting from middle of the index finger and middle finger signifies that the person will be kindhearted and jovial. He/she will have good humor and one can easily identify them.

A twist heart line starting from middle of the index finger and middle finger as the figure 1.1 show above. The person is timid but looks strong. The person will be humor but do not express directly. He always maintains silence and just ignores others. People feel him as dangerous but he is actually too good.

A heart line starting from Middle finger signifies that the person is not good will have cruel thoughts. They are very dangerous and have criminal ideas. They cheat others. They sometimes lacks back getting self respect. This person will have heart line starting from Saturn mount.

A heart line starting from middle finger but not from above but starts from center. The person will have stupid ideas; they look innocent but are cunning. They cannot be judged easily. They talk like Brave but are cowards. They can easily hide their identity. They hide secrets. If the heart line has twists the person is timid but hide himself as brave.

Hear line starting from Sun line signifies many health problems they will unnecessarily affects from diseases. They cannot overcome all problems in their life. This line is seen very rare.

Absence of heart line signifies that the person is absence of nature. His nature, care and health is decided by God.

HEAD LINE

A head line explains one's hard work and intelligence. Head line plays very important role in one's life. If no head line no creativity. The longer the head line the higher the creativity.

HEAD LINE JOINED WITH THE LIFE LINE

This head line actually defines that the person is completely attached to his family relations. He ends his entire life with relations. They are little sensitive. This is not so lucky mark to have.

If the line ends in the middle: it signifies that the person will not show interest on building his future. He wastes his time. We can take him as an average student.

If the line ends at the edge of the palm: this signifies that the person will be intelligent and topper. He cannot reach what he wanted because of the bounded to family.

A slope head line : this is one of the good mark to have, the people having this line are hard workers and can achieve only by material success. He is not so lucky. He do not have too general knowledge. He do not risk for new thoughts.

HEAD LINE SEPARATED FROM LIFE LINE

This head line is one of the most important lines to have. The people with high position will have this type of lines. The person having this line give time to think differently. Will have independent thoughts. They do not depend on others. They depend on their own thoughts and ideas. They create their own future. They do not bother more about their relations. They can achieve anything in their life. This is the perfect line one should have.

If this head line ends in the middle of the palm it signifies that the person will find his own career and do not find interest in studies. However he can finally achieve success.

If head line ends at the end of palm it signifies that the person is very creative and intelligent. He can find his career in high position fields like astronomy, research institutes etc.,

A fork at the end of a straight head line but ends in the middle of palm signifies that the person is very creative but he cannot get into his career because of opposition from his family or others.

A fork at the end of a straight head line ends at the end of palm signifies that the person is creative and artistic. Writers, painters, wildlife photographers, Graffitist, cartoonist etc., will come under this category. Most of creative persons will have this line.

An island on the head line indicates that the person strives very hard to grow up or very weak in analyzing problems.

A dot on the head line indicates that the person will suddenly stops thinking or he may suffer from any problems. This varies from days to months.

Cross on the head line indicates that the person will spoil his entire career. He may commit suicide. A cross on the head line is the perfect indication of suicide. Cross also indicates that the person will have head injury or sometimes coma.

Q: Which head line is perfect to have?

A head line should be straight and little sloppy. It should not have any cuts or breaks. It should be very clear and dark. It should touch the end of palm with a fork. This head line is the perfect of all the headlines.

LIFE LINE

A line which gives us the information of threats
and life span is called as life line. The life line is
one of the important lines to have. Life line
should be long and curved; it should reach the
bracelets of life.

LIFE LINE AND VARIATIONS

If the life line is starting from edge of thumb
finger indicates that the person fears of
everything and panic unnecessarily.

If the life line starts from Jupiter mount then
the person will be in high position. A political
leader, Head of project will come under this
category and they will have such lines or if the
branch of life line reaching Jupiter mount.

If the life line is full it means ends at the
bracelet of life then the person will enjoy full life
span, if it has some breaks in the middle he may
fall illness unnecessarily.

If the fate line is incomplete or ends in the middle then the person cannot have full life span he may live shorter life.

If the life line starts from the mount of Mars then the person will have good health, he may not suffer from any diseases.

Islands, dots, breaks, cross on the life line are not good to have they indicates threat at the particular age.

MARKS ON THE PALM AND THEIR SIGNIFICANCES

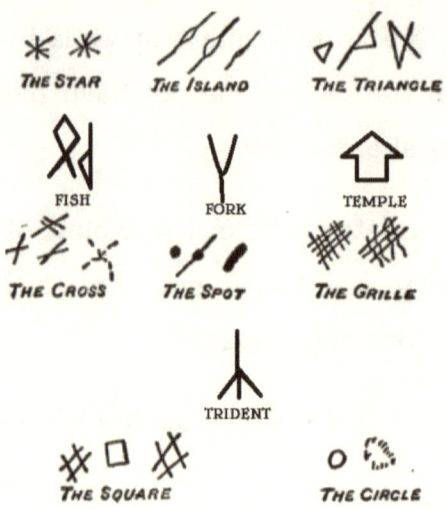

THE STAR   THE ISLAND   THE TRIANGLE

FISH   FORK   TEMPLE

THE CROSS   THE SPOT   THE GRILLE

TRIDENT

THE SQUARE   THE CIRCLE

The Marks on the palm may be Star, Square, Fish, Cross, island, fork, Trident, Branch, Circle, Griddle and Triangle.

STAR :

A Star actually refers to a good one. But it should not present at any undefined positions.

A star at the middle of fate line represents that the person is likely going to have some financial problems at that particular time.

A star at the end of fate line represents that the person will be famous in the society he will acquire world fame.

A star between head and heart line represents that the person is a great philosopher and can change thousands of minds.

A star on the middle or any part of head line represents that the person will commit suicide. There is possibility of a head injury.

A star on the sun line does not represents good. If may eat your name in the society so be careful.

A star on the Jupiter mount represents that the person will have a celebrity wife.

A star on the mount of mars represents that the person will be braver.

SQUARE

A square usually represents the protect from danger or financial loss.

A square near the fate line represents that the person will escape from financial problems or debts he have.

A square at the middle of head and heart line represents that the person is a famous astrologer.

A square near the life line represents protect from injury or any health problems or any road accidents.

FISH

A fish usually represent a sign of honor and wealth. A fish is a rare sign seen only at the end of life line. it also sometimes present near the mount of Saturn.

A fish at the end of life line represents that the person will have some good wealth and he will be happy even in his old age.

CROSS

A cross usually represents a sign of danger or threat. It should not present. It sometime also signifies good.

A cross near the sun mount represent that the person cannot get good name in the society or he feel unsatisfied for his career.

A cross near the plain mars or center of palm represents that the person will not have expected relations. He usually stay away from family.

A cross at the Jupiter mount represent that the person will have a beautiful, educated and understandable partner. She/he enjoy their partners company. A happy marriage couple will have this line.

A cross on the head, life, or heart line represents threat or any health operations.

A small cross near the bracelet of life represent that the person will feel uneasy while travelling.

ISLAND

A island usually represent a sign of loss.

Island usually present in the fate line. It represents that loss in business or any family health problems

Island at the end of fate line represent that the person family will find difficult to get ones prosperity if the person dies.

Simply inheritance problems.

FORK

A strange mark in palm. this do not present usually. So I cannot explain about this. This is a sign of magic arising in one's life.

A fork present at the end of Jupiter mount represent that the person life goes like a heavy growth chart. People will be in his way.

TRIDENT

A trident usually represents a good earnings. If it present at beginning of fate line, it represents that the person will have good earnings from multiple sources. It likes a rocket booster.

BRANCH

A branch is a position of support or help. If a branch of fate line is arising from Venus mount and another branch of fate line is arising from moon mount the person will become very rich and visit many foreign countries.

CIRCLE

A circle usually represents a sign of danger or unexpected situation. If a circle is present near the mount of mars the person will get problem from fire.

A circle is present near the head line slightly above and below the heart line but at the end represents that the person will have water problems.

A circle present at the sun line represent a grow in one's life.

GRIDDLE

A griddle usually represents a tensions, stress, over work or not being lucky. This is not a good line to have.

TRIANGLE

A triangle usually represents a sign of lucky. It helps one's life. If a triangle line is formed near the fate line, the person will save money for his future. Or he will have many properties and houses.

TEMPLE

A temple is a sign of luck. If a temple is seen on any part of the palm the person can be very lucky. Even without hard work the person can become rich.

SOME UNDEFINED LINES ON THE PALM

RING OF SATURN

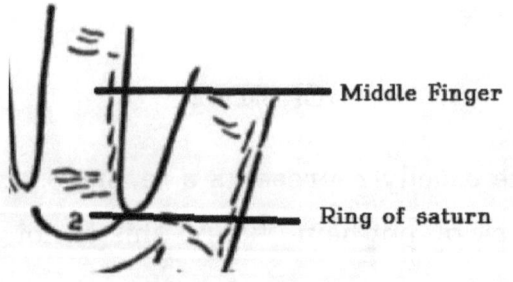

This is present near the middle finger arising like a semi circular mark. This indicates bad. No people cannot success in their life if they have this mark. Even though the work is easy they must do it again and again to accomplish.

After my complete experience I have found a remedy for this. The person having this mark can just walk three steps back while they start their work. Many of my members experienced good after I have suggested this. Am still searching how this works.

RING OF SOLOMON

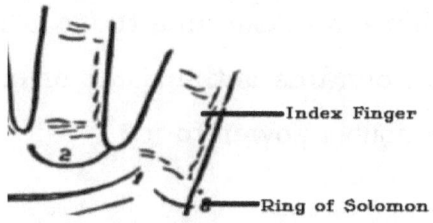

The ring of Solomon represent that the person will be optimist. He can easily expect what they are going to become in future. The ring of Solomon is actually the honor that the person will receive any precious awards in his life.

SUN LINE

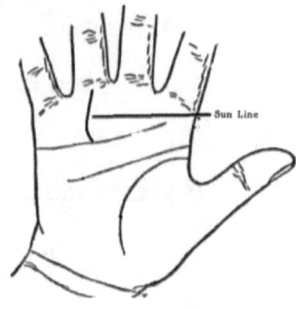

The Sun line is the main important line to have. Without sun line no one can raise themselves in society. If there is no sun line there is no achievement. Sun line is the sister of fate line which gives double power to it.

Complete palm analysis in palmistry

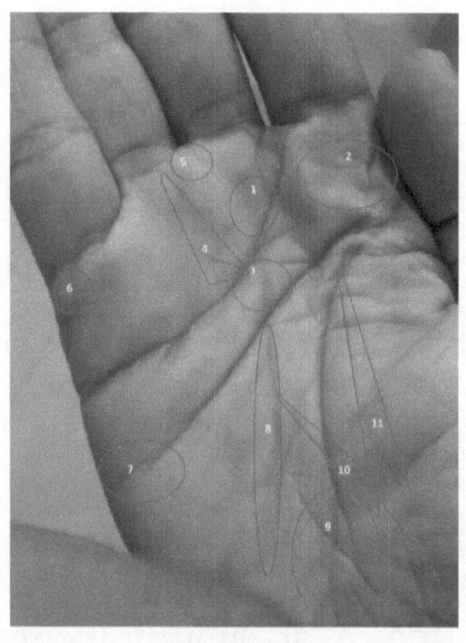

here am going to explain about every line in the palmistry, you can see many different lines that are not seen in usual palm, Here am going to explain about each and every line as shown below.

1. The mystery fish sign near Saturn mount
2. The tuning fork crosses near mount of Jupiter
3. The mystic star at the end of fate line
4. The mystic sun line near mount of sun
5. The double creative line between middle fingers

6. The healthy child line near mercury below little finger

7. The fork head line

8. The perfect fate line starting from mount of moon

9. The cross of ketu and **matsya rekha(fish line)** at end of life line

10. The branch of fate line arising from mount of Venus

11. The Venus line joining the cross of Jupiter

the Venus line joining at the end of Jupiter line or at Jupiter mount

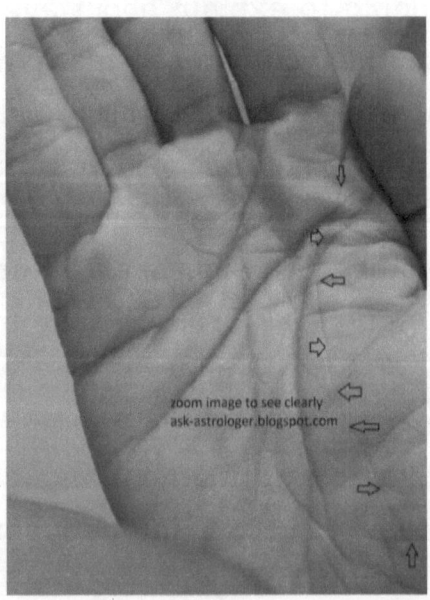

The line of Venus joining at the end of
Jupiter
mount represent that the person is occult
who can expect his future and can analyze
other. He will not troubled with any health
problems, he can impress anyone.

The cross of ketu and matsya rekha at the end of life line

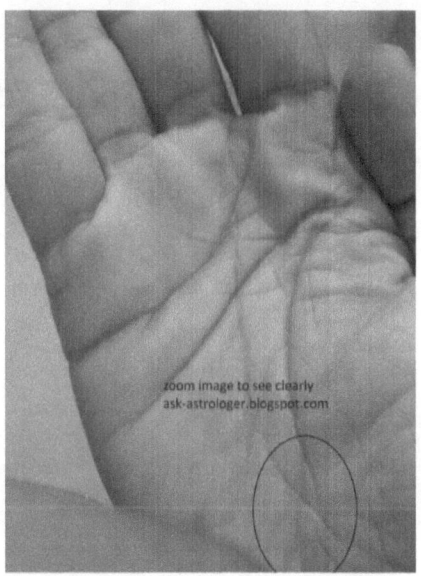

This is the most important line found in rich

people. if there is a cross found at the end of life line it represents that the person will be very happy even in his old age and a matsya rekha or sign of fish at the end of life line represent that the person will become more rich and wealthy.

if both marks are found and ketu mount touching fate line branch , no doubt it represent that the person is crorepati. What a lucky sign. The person will adore the god

The fate line starting from mount of moon and mount of Venus

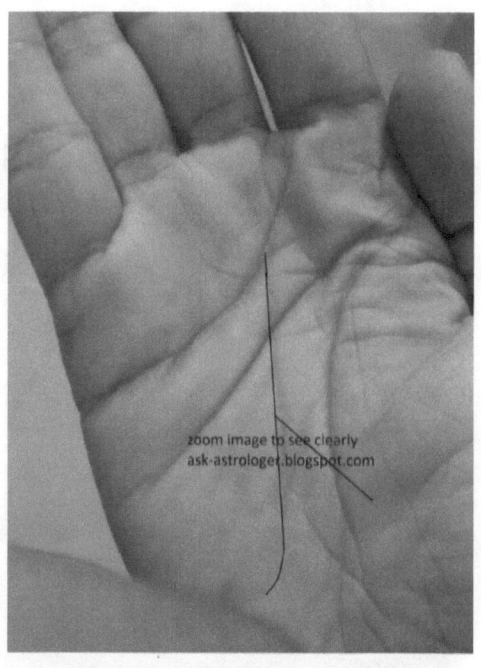

zoom image to see clearly
ask-astrologer.blogspot.com

If the fate line starting from mount of moon it represents that the person will have support from other especially from life partner. And if the fate line starting from Venus mount it represents that the person will earn huge income with little effort and he usually will have independent life.

if the fate line starting from mount of moon and one of the branch crossing Venus mount it represent that lucky will be in his side and become famous.

the fork at the end of head line

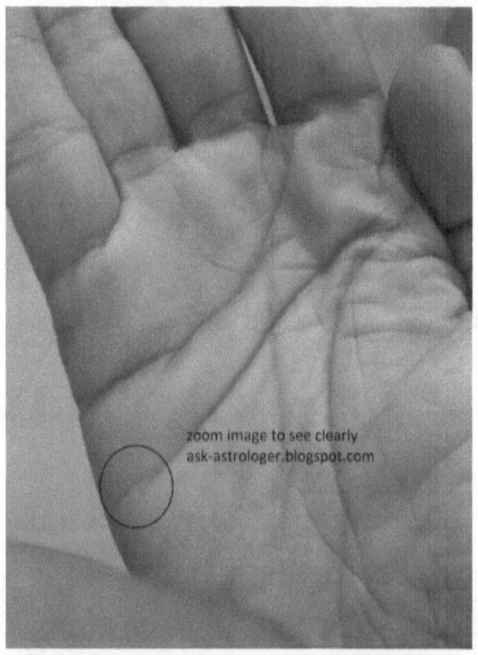

The fork at the end of fate line usually represent that the person is artistic and will have high creative ideas. if the head line touches moon mount with fork at its end represent that the person will become successful writer or novelist and acquire high name on aboard but not more in his native place.

Photographers, writers, actors, drama artist, painters, animators will come under this category

The healthy child line near mercury mount above marriage line

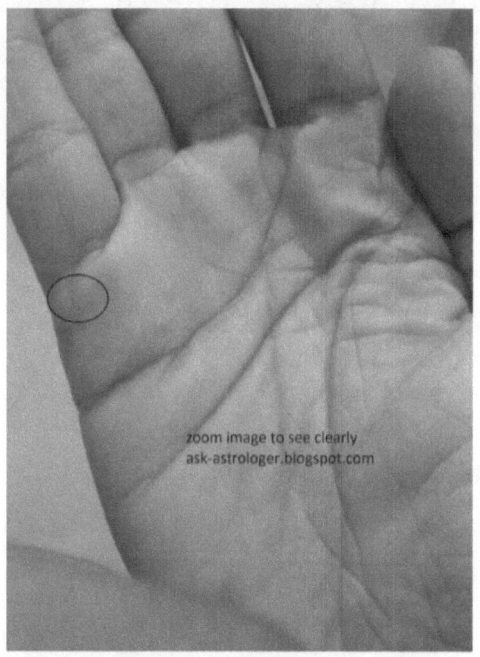

The thick and clearly visible line on the mercury mount vertically above the marriage line represent that the child will

be healthy. The dark line represent here is male child who will not be under the father control and the father faces some trouble. The child will be mischievous, healthy, intelligent, and handsome.

The double creative line between middle finger or line of education

The line present between the middle finger usually represent the line of study or education line. if they are two education lines that the person will have full grip on any subjects and will become famous. Here the first line is shorter that second it represents that the person will settle in his field and study as a habit or part time but finally become successful. the person will solve any difficult problems easily and he is also a philosopher

The sun line on the mount of sun in the palmistry

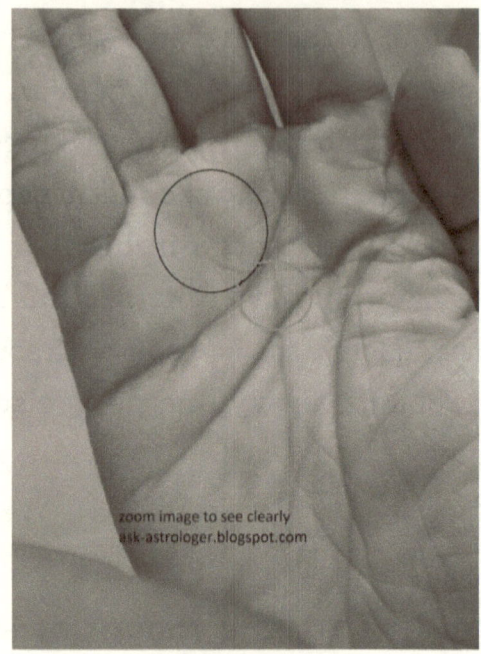

zoom image to see clearly
ask-astrologer.blogspot.com

see the black circle

The sun line usually represents the sign of
recognition without this line there will be no
name or the person will not prove his name
in the society, this is one of the most
important mark found in the palm near sun
mount. if it ends at heart line the person will
become success with his own efforts, if it
ends at head line he will guide and be on his
ideals and finally acquire success with hard
work.

The mystery star mark at the end of fate line

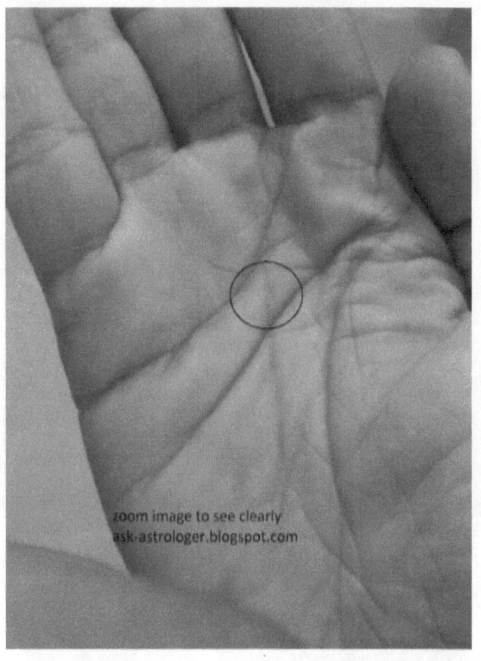

This is one of the rare and most important marks found in palmistry. A star on the palm represents problems but a star present at the end of fate line represents that the person will be more famous and success in his life. If it ends near heart line it represents that the person will be richer at his age 58. Here the clear fate line started from little about moon mount it represents that the person will start earning from 20 years old till 58. Wow!!

The tuning fork crosses near mount of Jupiter

zoom image to see clearly
ask-astrologer.blogspot.com

The tuning fork or a cross on mount of Jupiter usually represent happy married life, the cross usually represents that the wife/husband will be more understandable and romantic life. A tuning fork crosses represent a sign that the partner will be highly educated and wealthy will be beautiful/handsome and will be more understandable.

The mystery fish sign near Saturn mount on palm

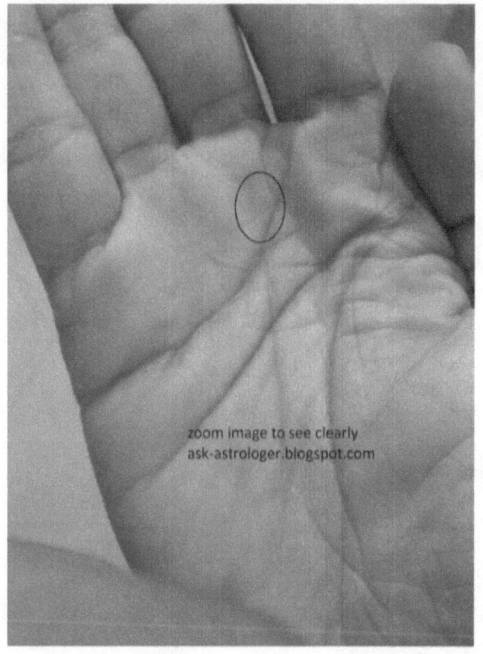

zoom image to see clearly
ask-astrologer.blogspot.com

This is the one of the rare and important mark seen in the palm, this line usually do not found on every hand. The sign of fish on

Saturn mount represents that the person will have a great death, world remember him. I have found this strange mark on one of my friends.

1. Which hand to analyze in palmistry left or right?

When come to question which hand should analyze for boys and which for girls, it is actually said that right hand for boys and left for girls in Hindu astrology, but here is few more to know, actually you can see complete future what happens if you neglect your life, the right hand denotes your current life and left hand denotes mirror palm. you can see three main lines head, heart and life similar mostly. if your right palmistry is good it means you are in right situation, if the right

hand is bad as left it means you are
neglecting your future. if two hands are bad
then you are not at all caring your future, as
science knows palm lines changes
periodically. so just check your hand and
analyze how to follow your life.

How to calculate age of marriage in palmistry.

This is one of the most arising question for
girls who's marriage is not getting early,
here i will explain how to calculate our
marriage year.

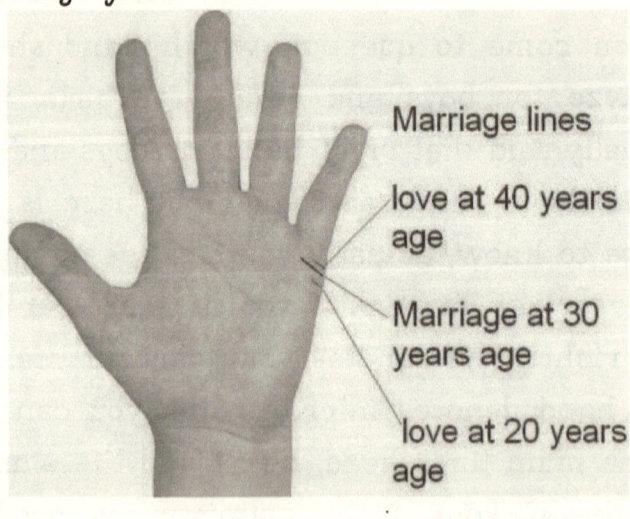

Marriage lines

love at 40 years age

Marriage at 30 years age

love at 20 years age

First look at your palm the lines below the little finger on mount of mercury are called marriage lines. Now see the number of lines. starting from heart line to end of little finger is called marriage age prediction, taking from heart to end of little finger draw a line, the complete line note to be 55 years, now starting from heart line see the marriage line which is longer of all draw a point on it, if the line is at the center note it as 28 years, and so on you can estimate the marriage age.

Why do we feel sometimes this incident has happened early?

This is one of the most mysterious question, we sometime feel I think it has happened early, we actually feel amazed when such incident happens. When come to clarity, this is not easy to believe-as we know human brain is most powerful in universe and no animal have such power, we know few animals can analyze deep sounds example

Elephants can hear sounds where humans can't here we fail to be powerful near elephants. Another example an Eagle can examine a rat running on the ground even it is in such a long distance in the sky because of its eye power, here also human brain fails to stand. But you already know baby child has ability to swim it is inbuilt act when they born, but while growing babies lost their act of swimming, in the same way the babies up to the 1.5 years can analyze future first itself because they are not so broad minded they forget early. When come to finally human brains have nerves which can analyze future but they are not active, they just get activated very rare, if we study to analyze I can say confidently one day human can analyze future.

Is our future decided at first itself, can we see our future

This is one of the most arising questions for

everyone is our future decided by god early, it is exactly false, our future is not decided early itself, it is decided by our ruling planets. Each planet ruling person has their own life decision for example if you're ruling planet is Jupiter and the nakshatra is Mula your life will be decided at the age of 7. In this age the persons ruling planet will analyze the persons behavior, mostly thinking for what, or dreaming for what, if the child dreaming or love to play, then he will surely become a person in the field of games.

Which type of persons can attract girls easily?

In palmistry and astrology which boy attracts girls easily are one whose life is mostly influenced by planet Venus. if you found more than 3 vertically line moving upwards on mount of Venus are said to be most romantic and he can attract opposite sex easily. In practical the boys who born in the months of January, July are said to be easily attract girls

What happens if you get bald head at teen age or early?

Most of the boys worry if they are getting their hair off and starting bald age at teen age it does not mean that they are getting old early. when come to clarity the boys who are getting bald at age of 16 are said to be most genius persons, because we already know when human was in the starting formation(early man) they had hair on all the body, when they are started updating all the hair wiped off and left only on the head. This is because of climate conditions and mainly the improvement in knowledge. The main reason for the bald head is the nerves in brain which have few voltage, when they are having more power they don't care of hair and the hair get off because of such more power. if the bald head starts from age of 35 it is because of mental problems and over thinking, so there is lot of difference between the teen bald head and normal bald head.

You may question me, then why do girls have full hair are they are not intelligent. This is silly question because human body has three systems as nervous, digest and sexual reproductive system, for girls the reproductive system plays a strong role that nervous because the hair is not bald but they are intelligent...

In future as our body has lost most hair, the male may lose their hair on head too, this could happen soon, as saying technology kills health..
Note most of girls in now do not care of hair (LOL)

Which girls will fell in love fast

This is one of the most arising questions for boys that which girls will fell in love fast. I will give you a small explanation this is just out for fun am not responsible for the post. When come to the question which girls will fell in love fast. I will give some points

1. Actually girls whose skin is honey or white in color will fall in love fast than girls with dark color faces.
2. Most of the girls fell in love when the boy is funny, humor or does idiotic things; if you are fun then you can attract girls easily.

3. When come to again another hint, the girls whose hair is long will fell in love faster than the girls with shorter or thin hairs.

4. Most of the girls fell in love with boys if they are innocent, funny, make silly things.

This book provided with basic palmistry, but the subject is still vast to know. I will explain about the complete details of mounts, diseases, careers, Raj Yog, remedies, months, stars and more in my next book

"THE MYSTERIOUS PALMISTRY"

@@@@@

Yours Author

Ravi R Naik

(Buy the NEW D Street Novel a suspense thriller)